# Discovering the Secrets of the Sea

by Elizabeth J. Natelson

Copyright © by Harcourt, Inc.

All rights reserved. No part of this publication may be reproduced or transmitted in any form or by any means, electronic or mechanical, including photocopy, recording, or any information storage and retrieval system, without permission in writing from the publisher.

Requests for permission to make copies of any part of the work should be addressed to School Permissions and Copyrights, Harcourt, Inc., 6277 Sea Harbor Drive, Orlando, Florida 32887-6777. Fax: 407-345-2418.

HARCOURT and the Harcourt Logo are trademarks of Harcourt, Inc., registered in the United States of America and/or other jurisdictions.

Printed in Mexico

ISBN 978-0-15-362479-7
ISBN 0-15-362479-5

2 3 4 5 6 7 8 9 10   126   16 15 14 13 12 11 10 09 08

Harcourt
SCHOOL PUBLISHERS

Visit The Learning Site!
www.harcourtschool.com

## Introduction

When you look at a globe, you see more water than land. In fact, 71 percent of Earth's surface is ocean. The ocean averages 5000 meters (16,000 ft) deep. Sea organisms living in all of that water are very different from creatures we find on land. And sea organisms don't live on just the ocean's surface. They live all through the water and even at the bottom of the sea.

If you were a scientist exploring the ocean, what would you see? Oceanographers and marine biologists study many things, such as plants, animals, underwater volcanoes, and waves. They scuba dive near the surface and guide submarines down to the depths.

A clownfish in a sea anemone

People are land organisms, so it's a special challenge to do research under water. Scientists invent special equipment to help them explore the deepest parts of the ocean. Some of what they learn about the ocean and its organisms is truly amazing.

## Clownfish

If you were a marine biologist, you would learn about sea organisms. You might study the way clownfish and sea anemones help keep each other alive.

Clownfish are small and defenseless. Even when they are fully grown, they are only about 13 cm (5 in) long. If a clownfish swims around in a reef, it is in great danger of being eaten by a larger fish or another organism. It needs a safe place to hide.

Anemones look like flowers, but they are actually a kind of animal called a polyp. A polyp has a soft, simple body shaped like a bag that is open at one end, with tentacles around the opening.

A sea anemone is not as helpless as a clownfish. Its tentacles have stingers that can paralyze the small animals it feeds on. Those tentacles also stop most organisms from attacking it. Clownfish, however, can swim unharmed among the anemone's tentacles. The anemone's venomous stingers don't bother them. By staying among the stinging tentacles, the clownfish are safe from predators.

Sea anemones, on the other hand, can be harmed by butterfly fish. Butterfly fish grow to about 20 cm (8 in) long. They can destroy an unprotected anemone. An anemone's defenders are, surprisingly, the little clownfish. Of the 1000 species of sea anemones, 10 species form partnerships with clownfish. When several clownfish huddle among an anemone's tentacles, the butterfly fish stay away.

A clownfish hiding in a sea anemone does not grow larger until there is space. When a bigger fish dies, the smaller and younger ones suddenly grow larger.

# Coral Reefs

If you went scuba diving to study a coral reef, you would see a wonderland of shapes and colors. The reef itself is a rough ridge built from the hard body coverings of corals and other organisms. Hundreds of animals grow on the reef, hide in it, hunt around it, and swim past it. You might see crabs, sharks, jellyfish, and sponges.

You would certainly see coral animals. One end of a coral polyp's body makes calcium. The calcium builds up into a stony cup around the polyp and is attached to the reef. The cup is large enough for the polyp to completely hide inside.

At the other end of its body, the polyp has an opening. It uses this opening for food coming in and waste going out. Around this opening the polyp has either six or eight tentacles, depending on the type of coral. The polyp uses its tentacles to sting and paralyze passing prey and then to reach out and grab its victims.

The corals that build coral reefs must live near the water's surface because important algae live inside the corals, and algae need sunlight.

**A coral reef**

Algae and corals help each other stay alive. Using sunlight for photosynthesis, algae provide a coral with carbon and energy. The coral provides a home for the algae and may give them nutrients from the animals it eats.

Reef-building corals live in warm, shallow water, such as the Caribbean Sea. However, other species of coral can live at great depths. Scientists have seen coral in the cold darkness at almost 6000 m (nearly 20,000 ft) below the ocean's surface.

How can a great coral reef develop from a polyp only 1 to 3 mm (0.04 to 0.12 in) across? The polyp's calcium cup is attached to the reef. When the animal dies, its cup stays there. New corals build their cups on top of the old ones. There are many polyps living in a reef, and over time their cups build up the great reef.

Because of the way it forms, the reef has many caves and crevices where predators lurk and their prey hide. Every organism in the sea has its own special way to live and survive.

## The Octopus

Another organism that has tentacles is the octopus. Marine biologists study the many different types of octopods. One small species of octopus grows to only 5 cm (2 in) long, but a giant octopus reaches 5.4 m (18 ft) long, and its tentacle arms can stretch 9 m (30 ft) across.

Octopus

Regardless of their size, all octopods have some things in common. They don't have bones or shells. In fact, their bodies are so soft that they can slip into narrow cracks to hide from enemies and watch for prey.

An octopus most often eats crabs or lobsters, even though these animals can fight back with pinching claws. With two rows of suckers running along each of its eight arms, an octopus firmly grabs a lobster and pulls its prey close to its mouth. As the octopus bites down, poison from its saliva paralyzes or kills the lobster. The blue-ringed octopus has venom strong enough to kill an adult human.

Lobsters move around mostly at night and stay on the ocean floor. How do octopods find their food? Octopus eyes are much like those of a vertebrate (an animal with a backbone), so they have very good eyesight.

Octopods are also intelligent. No other boneless animal has as large or as well-developed a brain. Biologists have found that octopods can solve problems and later remember what they have learned. Intelligence makes them good hunters.

They are emotional as well as smart, and they change color with how they feel. When angry or afraid, they can switch from pinkish to brown and back again, once or many times.

An octopus is not only a predator—sometimes it is prey. To get away from predators, an octopus shoots a dark liquid ink at its attacker. The cloud of ink hides the escaping octopus, and the ink of some octopods paralyzes the sensing abilities of a predator.

Hidden by its ink cloud, the octopus doesn't just crawl away. It darts off swiftly by shooting water out behind it. Then it slips into a hole or crack until danger is gone and its next meal comes into sight.

A female octopus uses a den to care for her eggs. In two weeks' time, she lays about 150,000 eggs. Then she spends 50 days guarding them. She even spurts water on them so that they stay clean. After they hatch, the young look like tiny versions of their parents. They float on the ocean's surface for several weeks, drifting with the current, and then live on the sea floor.

## Animals that Glow

Some scientists study *bioluminescence,* which is light produced by a living thing. (*Bio* means "life," and *lumin* means "light.") On Earth's surface, only fireflies and a few other organisms can make their own light. It's a different story in the top part of the ocean, where many creatures are bioluminescent.

**A female deep sea anglerfish**

8

Light can attract prey. A female deep sea anglerfish lurks in the darkness, 800 meters (2625 ft) below the ocean's surface. No sunlight ever reaches this depth. Like other females of her species, she draws prey with a snout that serves as a sort of fishing pole with millions of glowing bacteria on its tip. Small sea animals swim toward the light, and they end up being eaten.

Light can also turn a predator into prey. At the same ocean depth lives the sea cucumber. This boneless animal looks like a cucumber—except for the waving tentacles around its mouth. Its size ranges from 2 to 200 cm (0.75 in. to 6.5 ft) long. A sea cucumber's body is covered with a sticky, jellylike outer layer. This layer glows when it is touched. When a sea cucumber is attacked, the glowing skin comes off and sticks to the attacker, putting the attacker in danger of being seen by even larger predators.

Light can stop predators. The shining tubeshoulder is a fish that can shoot glowing slime at a predator. In the total darkness of the deep sea, the flash of light distracts a predator long enough for the shining tubeshoulder to escape.

Light may also be a warning. The bamboo coral shines when it is touched. Then it lets loose a gooey slime that gums up the gills of an attacking fish.

Light helps prey hide. When an animal looks up near the ocean's surface, it sees the brightness of sunlight on water. Any animal overhead stands out as a dark shape against the brightness. How can a sea creature hide from animals below it? It lights up.

The cockatoo squid, for example, has a clear body you could see through, except for its dark eyes and ink glands. Below its eyes are photophores, small organs that produce light. (The root word *photo* means "light.") The photophores always aim light down. From below, the cockatoo squid matches the brightness of the shiny surface and cannot be seen.

Many kinds of fish have photophores. Each of these special organs is small and simple, but one fish can have thousands of them. Some fish can make light with just their photophores. Others must either eat bioluminescent prey or be home for glowing bacteria.

# A Sea Within an Ocean

Some biologists study the connection between sea animals and other organisms. When Christopher Columbus was sailing across the Atlantic Ocean, he discovered an area of calm water and floating seaweed. He thought that he must be very close to land, but he wasn't. He had reached an area now called the Sargasso Sea, named for a brown algae called *Sargassum*.

The Sargasso Sea is surrounded by the Atlantic Ocean. The Atlantic's currents swing clockwise around it, leaving the Sargasso Sea lying quietly in the middle.

If you visited this area, you might first notice the brilliant blue of the water. You would also see brown clusters of *Sargassum*. On the algae's branches are little air sacs that look like yellow-brown berries or grapes. Those air sacs are natural floats that keep the branches from sinking.

For some kinds of eels, the Sargasso Sea is the place life begins and ends. The American eel, for example, is hatched as a larva in the Sargasso Sea. There it floats for 9 to 12 months until it is about 6 cm (2.4 in) long. It becomes a glass eel and moves to the Atlantic coast along the United States.

Growing and changing, the glass eel finally becomes a yellow eel in a freshwater river linked to the sea. Female eels may grow to be 1.5 m (5 ft) long, and males reach up to 61 cm (2 ft). At night they feed on many creatures, from insects to fish.

After five to twenty years, these eels change again. Their eyes and fins get bigger and their color changes. In the autumn night, they follow an inner call back to the Sargasso Sea. When they reach its warm, still waters, they lay their eggs, and then they die.

Sea turtles are very different. They are born on land, scurrying to the sea as soon as they hatch. The hatchlings don't stay near the coast because there are too many predators. Instead they swim to safer places, including the Sargasso Sea. They devour sea animals such as shrimp, snails, crabs, sponges, and jellyfish.

Slowly they grow larger. A green sea turtle may reach about a meter (3 ft) in length and about 140 kg (about 300 lb) in weight. Finally, after about 11 years, the turtles return to land to lay their eggs.

The algae of the Sargasso Sea provide shelter and nourishment for many different animals. In return, the *Sargassum* gain nutrients from the waste materials of all those animals.

**Sargasso Sea**

## A Deep-Sea Volcano

If you were a geologist, you might study underwater volcanoes. In the Pacific Ocean west of Oregon, the Axial volcano stands 1000 m (3300 ft) high, but the top of this volcano is 1400 m (4600 ft) underwater. Axial is so active that the U.S. government has a long-term research center near it.

To study Axial, scientists send down ROPOS, a remote-controlled submarine. ROPOS can shine lights through the total darkness of the deep sea, take video images, and pick up samples with mechanical arms. The samples are studied at the research center.

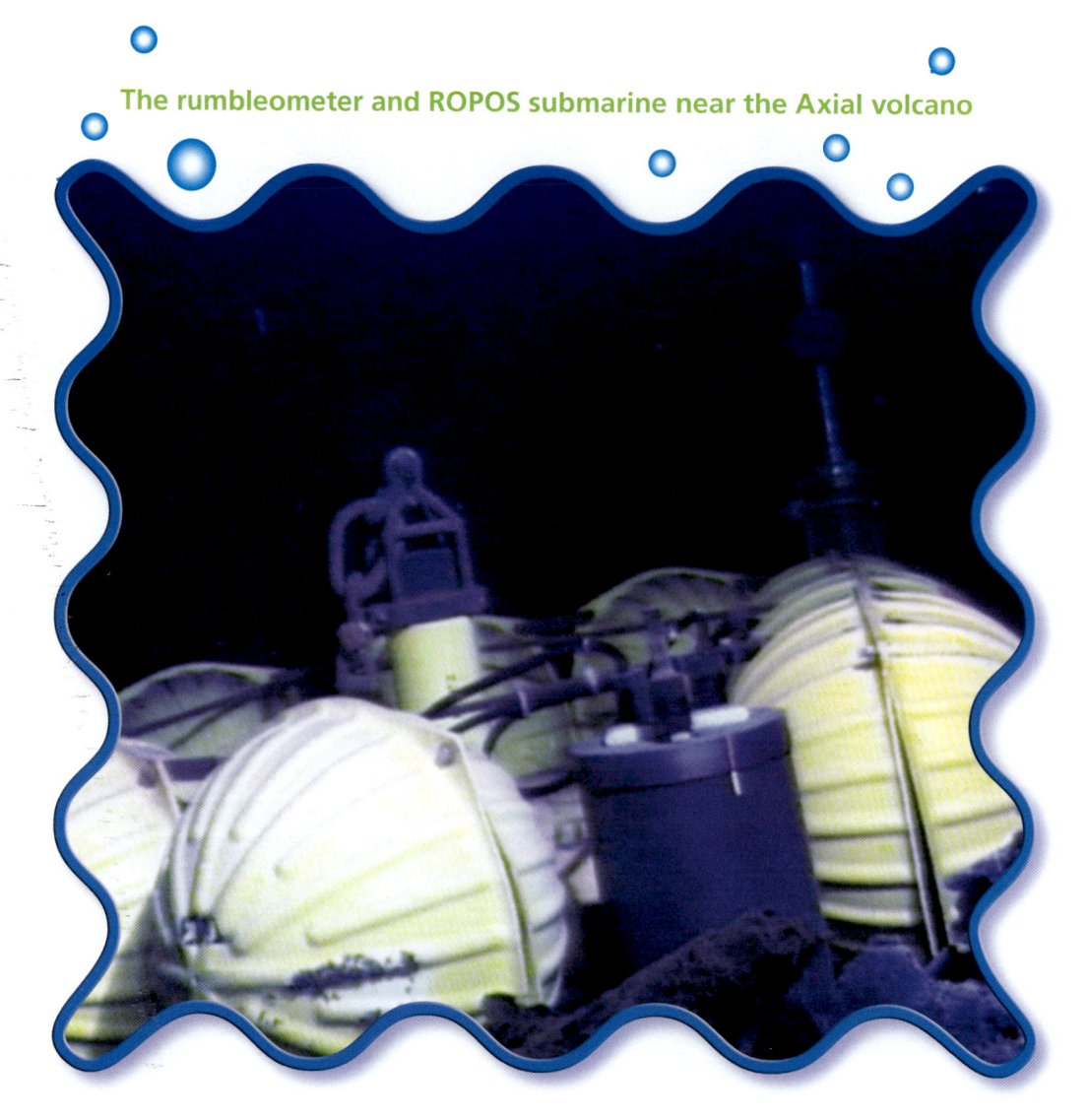

**The rumbleometer and ROPOS submarine near the Axial volcano**

When Axial erupted in 1998, earthquakes trembled and lava flowed. In fact, geologists later estimated that 200 million cubic meters (260 million cu yd) of magma moved, both under the sea floor and on it.

Before Axial erupted, scientists had placed a measuring instrument called a rumbleometer on the sea floor. A rumbleometer measures underwater pressure and temperature near a volcano. It can measure up-and-down movements because water pressure changes according to depth.

During the eruption, the rumbleometer was caught in a lava flow, lifted up 3 m (9 ft), and then slowly set down as most of the lava drained away. The rumbleometer became stuck in the cooled lava. Scientists found it, and it was still working. Icy seawater had cooled the lava before it could melt the instrument.

Axial volcano stands on top of a mid-ocean ridge. Under the sea, just as on land, Earth's surface is made of huge plates that move very slowly beside each other. In some places they bump together, and in other places they pull apart. Earth's long mid-ocean ridges are lines where Earth's plates are pulling apart. Near Axial, the seafloor is spreading apart at about 6 cm (2.34 in) each year.

Where the seafloor spreads, magma can move up from deeper in Earth. Often magma cools and forms new crust without breaking into the water. In other places it erupts through one of the thousands of undersea volcanoes on the mid-ocean ridges. It cools and becomes a new layer of crust.

Axial is not just an opening where magma leaks out. Axial is a seamount, which is a volcano standing by itself on the ridge floor.

Besides being on a mid-ocean ridge, Axial is also right over a hot spot. A hot spot is a magma channel flowing from deep in Earth to near the surface. Because of the hot spot, Axial has more magma and erupts more often than most volcanoes.

## Underwater Hot Springs

Near undersea volcanoes geologists have found hot springs, which they call hydrothermal vents. (*Hydro* means "water," and *thermal* means "heat.") In 1977, three scientists from Woods Hole Oceanographic Institute were aboard *Alvin*, a small deep-sea submarine. They were astonished to discover not only hydrothermal vents but also a community of living creatures all around the vents.

Near an undersea volcano, hot magma is trapped a short distance underground. When seawater trickles down to the hot zone, it gets heated up to about 300°C to 400°C (572°F–752°F). This water is so hot that metals dissolve in the water. When the superheated water spurts upward into the sea, it is rich with sulfur compounds. These compounds would be poisonous to land animals, but they support life for more than 300 species of vent creatures.

If you rode in *Alvin* down to a hydrothermal vent, your eyes would be drawn to the chimneys. These chimneys grow constantly. When hot water spurts into the cold ocean, it cools quickly. The sulfur compounds settle out and stack up. A chimney can pile up to more than 10 m (33 ft) high.

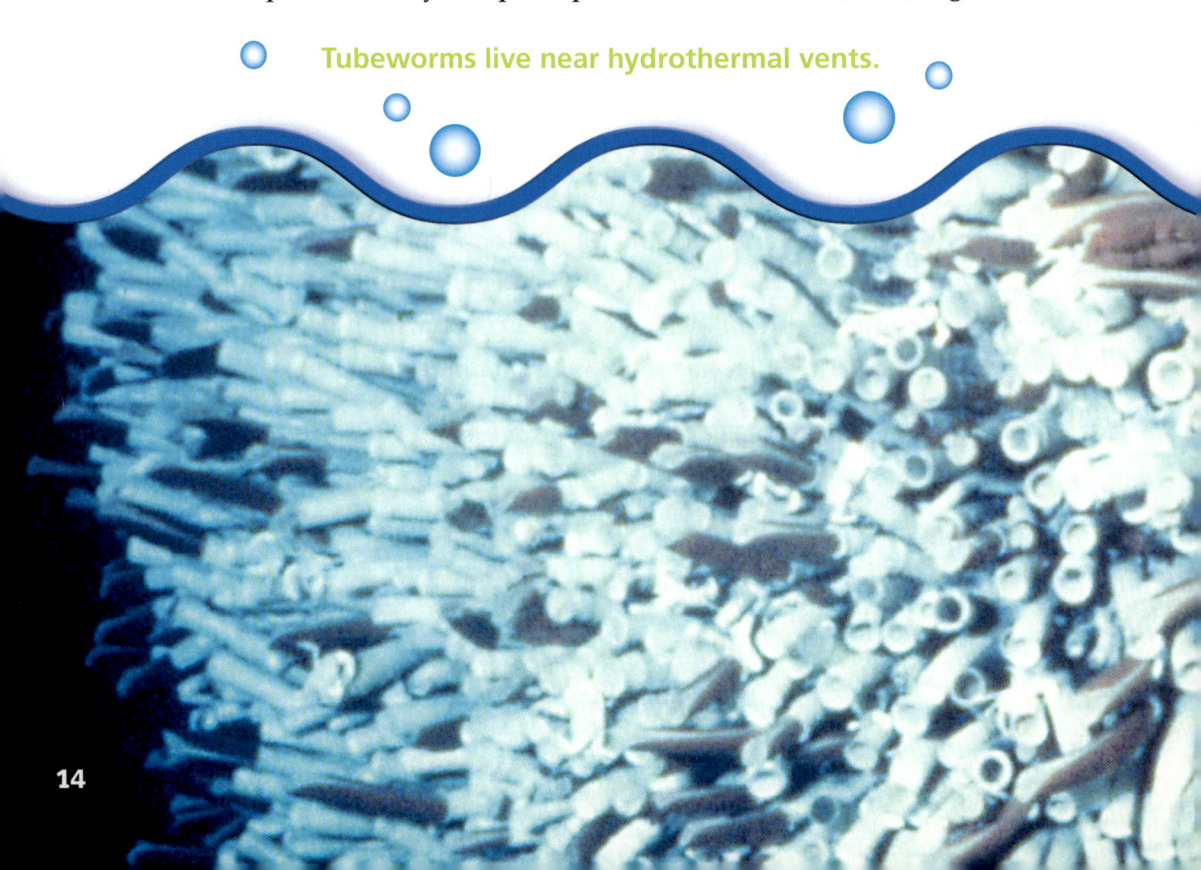

Tubeworms live near hydrothermal vents.

You might also see a black smoker, a chimney with black liquid coming out like smoke. The liquid is actually water mixed with sulfur compounds.

You would not be able to see the microbes (microscopic organisms) that make all other vent life possible. Remember that on Earth's surface, life depends on photosynthesis, which is the process by which plants use energy from the sun. In the endless night of the deep sea, microbes use chemosynthesis to get energy from sulfur compounds. Snails and other animals eat the microbes, and larger animals eat the snails. All of these organisms cluster closely around the hot, life-giving vent.

In the midst of all these organisms you would still notice the gently waving tubeworms. Hundreds or thousands of tubeworms clump together to form a colony. Tubeworms don't eat microbes or anything else. Tubeworms don't have mouths or stomachs. Instead, microbes living in the tubeworms' bodies nourish them directly.

How do scientists know what happens in a tubeworm? A group of scientists dove 3 km (2 mi) down in *Alvin*. Using *Alvin's* robotic arm, they carefully lowered a probe into a tubeworm to measure what was inside.

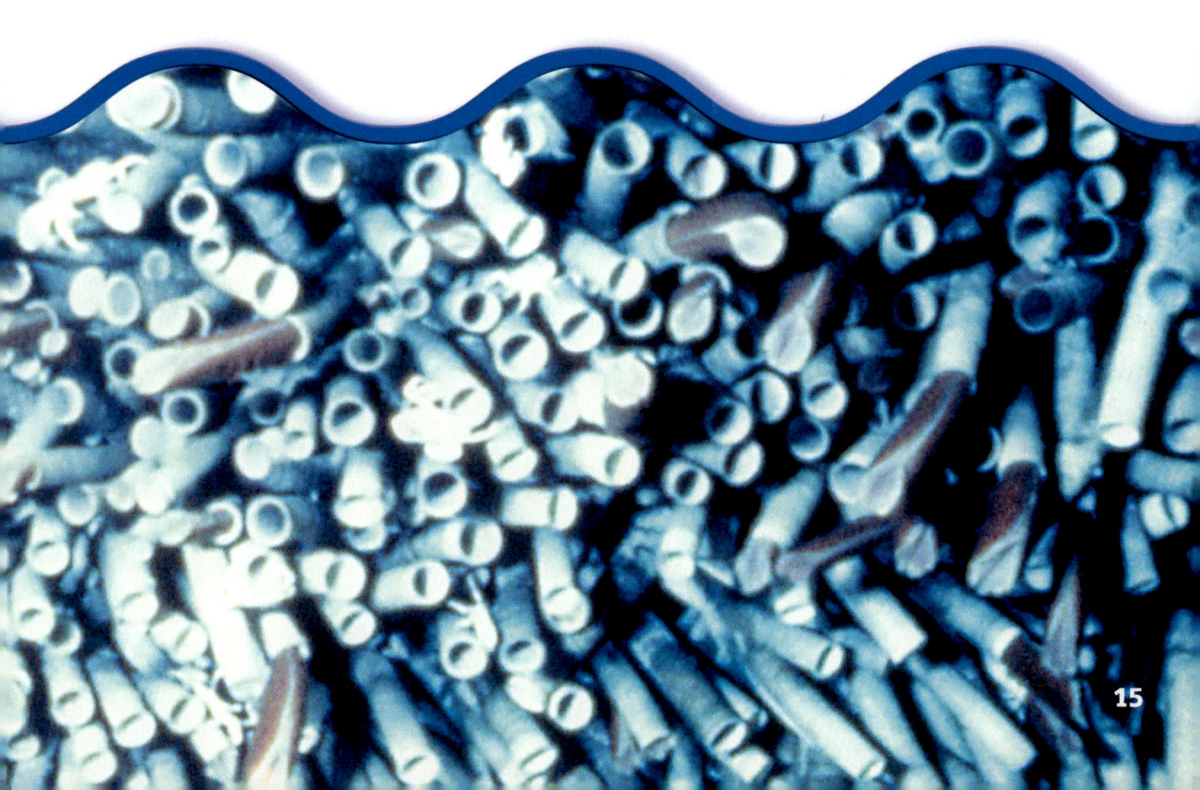

# Tsunamis

If you were an oceanographer you might also study tsunamis, and you could save lives by doing it. A tsunami is a gigantic ocean wave set off by an underwater shock, such as an earthquake or a volcano. Tsunamis are not caused by storms or tides.

In late 2004, a severe earthquake in Indonesia set off a tsunami that slammed into many Indian Ocean shores with waves up to 4 meters high. More than 300,000 people died.

Tsunamis are powerful because the force that causes them is immense. An ordinary windblown wave crosses the ocean's surface and does not affect the deep water. When a tsunami is started by an earthquake, however, even the deep water begins to move.

If you were in the middle of the sea, a tsunami could pass your boat and you probably would not notice it. It would seem to be part of the ocean's normal movement. Near shore, however, the sea becomes shallow. The wave piles up higher and higher. A towering wall of water rushes forward, destroying everything in its way.

Scientists cannot stop tsunamis, but they are trying to predict them. When instruments detect an undersea earthquake, government scientists measure it to determine whether it might cause a tsunami. If it might, they send out a warning, telling people to move quickly to higher ground farther inland.